JUN -- 2020

Jefferson Twp Public Library
1031 Weldon Road
Oak Ridge, N.J. 07438
phone: 973-208-6245
www.jeffersonlibrary.net

THE MONTHS OF THE YEAR

By EMMA CARLSON BERNE

Illustrated by TIM PALIN

Music by MARK OBLINGER

CANTATA LEARNING

Jefferson Twp. Public Library
1031 Weldon Road
Oak Ridge, NJ 07438
973-208-6244
www.jeffersonlibrary.net

WWW.CANTATALEARNING.COM

CANTATA LEARNING

Published by Cantata Learning
1710 Roe Crest Drive
North Mankato, MN 56003
www.cantatalearning.com

Copyright © 2020 Cantata Learning

All rights reserved. No part of this publication may be reproduced
in any form without written permission from the publisher.

Library of Congress Cataloging-in-Publication Data
Names: Berne, Emma Carlson, author. | Palin, Tim, illustrator. | Oblinger,
  Mark, composer.
Title: The months of the year / by Emma Carlson Berne ; illustrated by Tim
  Palin ; music by Mark Oblinger.
Description: North Mankato, MN : Cantata Learning, 2020. | Series: Patterns
  of time | Includes bibliographical references.
Identifiers: LCCN 2018053578 (print) | LCCN 2018056158 (ebook) | ISBN
  9781684104246 (eBook) | ISBN 9781684104093 (hardcover) | ISBN
  9781684104369 (pbk.)
Subjects: LCSH: Months--Juvenile literature. | Calendar--Juvenile literature.
Classification: LCC CE13 (ebook) | LCC CE13 .B475 2020 (print) | DDC
  529/.3--dc23
LC record available at https://lccn.loc.gov/2018053578

Book design and art direction: Tim Palin Creative
Editorial direction: Kellie M. Hultgren
Music direction: Elizabeth Draper
Music composed and produced by Mark Oblinger

Printed in the United States of America.
102019 002752

TIPS TO SUPPORT LITERACY AT HOME

WHY READING AND SINGING WITH YOUR CHILD IS SO IMPORTANT

Daily reading with your child leads to increased academic achievement. Music and songs, specifically rhyming songs, are a fun and easy way to build early literacy and language development. Music skills correlate significantly with both phonological awareness and reading development. Singing helps build vocabulary and speech development. And reading and appreciating music together is a wonderful way to strengthen your relationship.

READ AND SING EVERY DAY!

TIPS FOR USING CANTATA LEARNING BOOKS AND SONGS DURING YOUR DAILY STORY TIME

1. As you sing and read, point out the different words on the page that rhyme. Suggest other words that rhyme.
2. Memorize simple rhymes such as Itsy Bitsy Spider and sing them together. This encourages comprehension skills and early literacy skills.
3. Use the questions in the back of each book to guide your singing and storytelling.
4. Read the included sheet music with your child while you listen to the song. How do the music notes correlate to the words of the song?
5. Sing along on the go and at home. Access music by scanning the QR code on each Cantata book. You can also stream or download the music for free to your computer, smartphone, or mobile device.

Devoting time to daily reading shows that you are available for your child. Together, you are building language, literacy, and listening skills.

Have fun reading and singing!

What is a **month**? It's a way of measuring time! Each **year** has twelve months. They always go in the same order: January, February, March, April, May, June, July, August, September, October, November, and December.

December is the last month in the **calendar**. After December, we start over again with January. A new year has started. Read and sing along to learn more about the months of the year!

January begins a new year.
We play in the winter snow.

February is cold and gray.

In March, new flowers show.

January, February, March, April, May—
For June and July we cheer!

August and September, October, then November—
When December comes, the end is near.

January comes next!
That's the twelve months of the year.

Days get warmer when April arrives.
The grass and flowers come out.

Springtime ends in the month of May.
"School's over in June!" we shout.

January, February, March, April, May—
For June and July we cheer!

August and September, October, then November—
When December comes, the end is near.

January comes next!
That's the twelve months of the year.

In July and August, the weather is hot.
We love to swim in the pool.

September brings out colorful leaves.
School starts, and the air turns cool!

January, February, March, April, May—
For June and July we cheer!

August and September, October, then November—
When December comes, the end is near.

January comes next!
That's the twelve months of the year.

Leaves fall in October and November.
December brings winter days.

Then January starts a new year.
The months have ticked away!

January, February, March,
April, May—
For June and July we cheer!

August and September,
October, then November—
When December comes,
the end is near.

January comes next!
That's the twelve months of the year.

SONG LYRICS
The Months of the Year

January begins a new year.
We play in the winter snow.
February is cold and gray.
In March, new flowers show.

January, February, March, April, May—
For June and July we cheer!
August and September, October, then November—
When December comes, the end is near.
January comes next!
That's the twelve months of the year.

Days get warmer when April arrives.
The grass and flowers come out.
Springtime ends in the month of May.
"School's over in June!" we shout.

January, February, March, April, May—
For June and July we cheer!
August and September, October, then November—
When December comes, the end is near.
January comes next!
That's the twelve months of the year.

In July and August, the weather is hot.
We love to swim in the pool.
September brings out colorful leaves.
School starts, and the air turns cool!

January, February, March, April, May—
For June and July we cheer!
August and September, October, then November—
When December comes, the end is near.
January comes next!
That's the twelve months of the year.

Leaves fall in October and November.
December brings winter days.
Then January starts a new year.
The months have ticked away!

January, February, March, April, May—
For June and July we cheer!
August and September, October, then November—
When December comes, the end is near.
January comes next!
That's the twelve months of the year.

The Months of the Year

World
Mark Oblinger

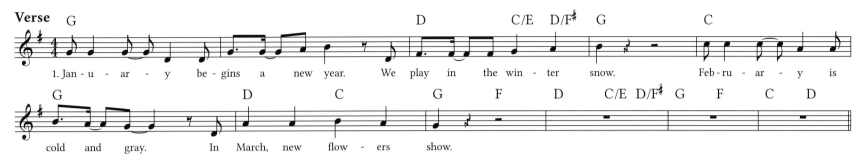

Verse 2
Days get warmer when April arrives.
The grass and flowers come out.
Springtime ends in the month of May.
"School's over in June!" we shout.

Chorus

Verse 3
In July and August, the weather is hot.
We love to swim in the pool.
September brings out colorful leaves.
School starts, and the air turns cool!

Chorus

Verse 4
Leaves fall in October and November.
December brings winter days.
Then January starts a new year.
The months have ticked away!

Chorus

GLOSSARY

calendar—a page or chart showing the months and days in a year

month—one of the twelve sections a year is divided into. The twelve months always come in the same order: January, February, March, April, May, June, July, August, September, October, November, and December.

year—a time period of twelve months

CRITICAL THINKING QUESTIONS

1. Give yourself a quiz. Cover your eyes with a blindfold. Can you name all twelve months of the year without looking at the book?

2. In what month is your birthday? Ask two friends or grown-ups in what month their birthday falls. If you have a calendar, write the birth dates down in the correct month.

3. Pick three months out of the year. What is one activity that you like to do in each month? Draw pictures of yourself doing those activities.

TO LEARN MORE

Clark, Claire. *The Calendar: How Long Is a Month?* North Mankato, MN: Capstone, 2012.

Sendak, Maurice. *Chicken Soup with Rice: A Book of Months.* New York: Harper Collins, 2018 (reprint).

Steffora, Tracey. *Measuring Time: Months of the Year.* Portsmouth, NH: Heinemann, 2011.